Cambridge Elements ≡

Elements in Geochemical Tracers in Earth System Science
Edited by
Timothy Lyons
University of California, Riverside
Alexandra Turchyn
University of Cambridge
Chris Reinhard
Georgia Institute of Technology

APPLICATION OF THALLIUM ISOTOPES

Tracking Marine Oxygenation Through Manganese Oxide Burial

Jeremy D. Owens
Florida State University
National High Magnetic Field Laboratory

CAMBRIDGE
UNIVERSITY PRESS

CAMBRIDGE
UNIVERSITY PRESS

University Printing House, Cambridge CB2 8BS, United Kingdom

One Liberty Plaza, 20th Floor, New York, NY 10006, USA

477 Williamstown Road, Port Melbourne, VIC 3207, Australia

314–321, 3rd Floor, Plot 3, Splendor Forum, Jasola District Centre,
New Delhi – 110025, India

79 Anson Road, #06–04/06, Singapore 079906

Cambridge University Press is part of the University of Cambridge.

It furthers the University's mission by disseminating knowledge in the pursuit of education, learning, and research at the highest international levels of excellence.

www.cambridge.org
Information on this title: www.cambridge.org/9781108723398
DOI: 10.1017/9781108688697

©Jeremy D. Owens 2019

First published 2019

A catalogue record for this publication is available from the British Library.

ISBN 978-1-108-72339-8 Paperback
ISSN 2515-7027 (online)
ISSN 2515-6454 (print)

Application of Thallium Isotopes

Tracking Marine Oxygenation Through Manganese Oxide Burial

Elements in Geochemical Tracers in Earth System Science

DOI: 10.1017/9781108688697
First published online: December 2019

Jeremy D. Owens
Florida State University and National High Magnetic Field Laboratory

Author for correspondence: Jeremy D. Owens, jdowens@fsu.edu

Abstract: Tracking initial ocean (de)oxygenation is critical to better constrain the coevolution of life and environment. Development of thallium isotopes has provided evidence to track the global manganese oxide burial that responds to early (de)oxygenation for short-term climate events. Modern oxic seawater thallium isotope values are recorded in organic-rich sediments deposited below an anoxic water column. An expansion of reducing conditions decrease manganese oxide burial and shift the seawater thallium isotope composition more positive. Recent work documents that thallium isotopes are perturbed prior to carbon isotope excursions, suggesting ocean deoxygenation is a precursor for increased organic carbon burial.

Keywords: deoxygenation, ε^{205}Tl, oceanic anoxic events, post-oxic, redox

ISBNs 9781108723398 (PB), 9781108688697 (OC)
ISSNs 2515-7027 (online), 2515-6454 (print)

Contents

1 Introduction

The reduction–oxidation state of the ocean–atmosphere system has undergone significant changes throughout Earth's history (Lyons et al., 2014). The broad-scale oxygenation history of the atmosphere is well documented (Berner, 2006; Lyons et al., 2014) and, generally, the oceans follow a similar pattern. However, there are numerous punctuated events throughout the geological record that document variability in the coupled ocean–atmosphere system (Saltzman and Thomas, 2012). Many of these events occurred in the Phanerozoic eon with associated biological crises, e.g., (mass) extinctions (Hannisdal and Peters, 2011). A wealth of geochemical proxies have been developed to potentially track past ocean conditions using sedimentary archives that transfer information from marine seawater to sediments (Tribovillard et al., 2006). Our geochemical redox toolkit has expanded greatly in the past decade with the utilization of Fe speciation, redox-sensitive trace elements, and isotopic analysis of many ele-ments from a variety of lithological archives (Tribovillard et al., 2006; Anbar and Rouxel, 2007; Raiswell et al., 2018).

Molecular oxygen is essential for many metabolic pathways and is important in regulating elemental cycles that are relevant to climatic and biological feed-backs (Anbar and Knoll, 2002; Berner, 2006). Relatively small increases in the geographical extent of reducing conditions can have profound effects on the availability of bioessential elements (i.e., Fe, Mo, V, and Zn; Scott et al., 2008; Reinhard et al., 2013; Owens et al., 2016) in the oceans, which in turn affects enzymatic processes (Anbar and Knoll, 2002; Saito et al., 2008; Glass et al., 2009). The sequence of chemical reactions/processes following oxygen deple-tion is important for marine organisms that can utilize various electron accep-tors to create free energy (Fig. 1A; Froelich et al., 1979).

Fingerprinting individual portions of the redox ladder or elemental pathways are ideal to reconstruct water column and sedimentary depositional conditions. In modern settings, this is possible (Canfield and Thamdrup, 2009), but owing to proxy specificity/limitations, burial diagenesis, and sedimentary transfer mechanisms, it is difficult to precisely fingerprint each process for ancient sedimentary rocks. Berner (1981) prescribed a general classification that sepa-rated the redox ladder into two zones: oxic and anoxic. The anoxic zone was delineated further into three categories: post-oxic (nonsulfidic), sulfidic, and methanic. Proxy development and application have led to the ability to accu-rately identify sulfidic conditions, with the means to also delineate sulfidic water-column (euxinic) from sulfidic sedimentary pore fluids (Scott and Lyons, 2012; Hardisty et al., 2018). However, unambiguously fingerprinting the initial global extent of deoxygenation or more broadly post-oxic conditions

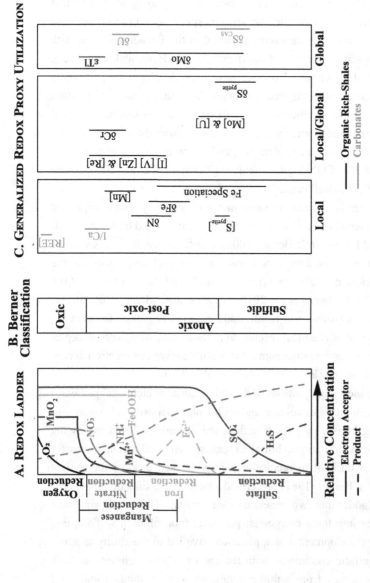

Figure 1 Conceptual figure that connects redox processes and potential proxy records. (A) Redox ladder adapted from Froelich et al. (1979), (B) Simplified redox classification modified from Berner (1981), which is related to (A). (C) Proxies that fingerprint portions of the redox ladder and are generally related to the redox scheme of (A) and (B). Several redox proxies are controlled by local, regional, and/or global processes that require supporting evidence for interpretation. The label color of εTl corresponds to manganese oxide reduction. Black lines in (C) represent proxies that have been developed or applied mostly to shales and the gray lines denote carbonate proxies. The line length indicates the approximate region in which the proxy responds to the corresponding redox conditions.

has been more difficult due to a lack of proxy specificity. One of the first minerals to respond to the earliest depletion of oxygen is manganese (Mn) oxides (Rue et al., 1997). However, tracking Mn concentrations only records the local process and requires investigating the global sedimentary record to reconstruct the response to ocean oxygen perturbations, which is not possible for the geological record. This is further complicated by the fact that Mn lacks multiple long-lived stable isotopes to generate an accurate mass balance framework.

Manganese oxides are an important marine sink and represent an oxic flux for many transition elements as they adsorb onto precipitated particles in the water column and on the seafloor (Hein et al., 2000 and references therein). For precipitation *and* long-term burial, Mn oxides require oxygen; as Mn oxides are one of the first minerals to dissolve under low oxygen conditions, they have the capacity to release adsorbed elements (Rue et al., 1997). The post-transition element thallium (Tl) readily adsorbs onto Mn oxides. Birnessite, a low-temperature Mn oxide mineral, imparts a large isotopic offset (fractionation) (Peacock and Moon, 2012). This Tl isotope fractionation represents the largest fractionation from seawater; thus it is the dominant control on the global Tl seawater signature in oxygenated oceans. The residence time of Tl is ~18.5 kyr (Fig. 2; Rehkämper and Nielsen, 2004; Baker et al., 2009; Nielsen et al., 2017; Owens et al., 2017), which is longer than ocean mixing times but is relatively short compared to that of other elements. The response to Mn oxides and its residence time makes it a promising element to track initial oxygen perturbations. Importantly, Mn oxides are likely perturbed during short-term climate events when the amount of oxygen in the oceans fluctuates, which has the potential to shift the global Tl isotope signature. This has been observed recently across two Phanerozoic oceanic anoxic events and one Precambrian oxygenation event.

2 Marine Elemental Cycle

Thallium concentrations and isotopes have been utilized for several decades (Shaw, 1952; Matthews and Riley, 1970; McGoldrick et al., 1979; Flegal and Patterson, 1985; Bidoglio et al., 1993; Rehkämper et al., 2002, 2004; Nielsen et al., 2004, 2005, 2006a; Rehkämper and Nielsen, 2004; Xiong, 2007), but the application of Tl isotopes as an ancient paleoceanographic tool is relatively recent (Nielsen et al., 2009, 2011b; Owens et al., 2017). This has been possible because of the numerous Tl isotope measurements of Earth materials (Rehkämper et al., 2002, 2004; Rehkämper and Nielsen, 2004; Nielsen et al., 2005, 2006b, 2017; Baker et al., 2009; Prytulak et al., 2013), which have provided a relatively well-constrained marine isotope mass balance (Rehkämper and Nielsen, 2004; Baker et al., 2009; Nielsen et al., 2017;

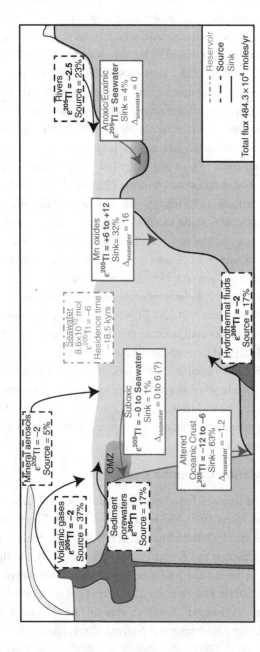

Figure 2 Marine Tl mass balance that includes the most updated data (Baker et al., 2009; Nielsen et al., 2009, 2011b, 2017; Owens et al., 2017). Black boxes represent the source fluxes with isotopic values (ε^{205}Tl) and the percentage of the total input flux to the ocean. Red boxes represent sinks from the ocean with isotopic values (ε^{205}Tl), percent of the total output flux, and the isotopic fractionation (the difference between ε^{205}Tl value and the seawater value labeled as Δ_{seawater}). Note that the averages for some of the fluxes are calculated by mass balance (see Owens et al., 2017). Total flux is 484.3×10^4 moles per year.

Owens et al., 2017b). Thallium has two isotopes, with atomic masses of 203 and 205, and their relative abundances represent ~30% and ~70% of the total, respectively. The isotopic ratios are reported relative to the NIST SRM 997 Tl standard:

$$\varepsilon^{205}Tl = 10,000 \times (^{205}Tl/^{203}Tl_{sample} - {}^{205}Tl/^{203}Tl_{NIST\ 997})/(^{205}Tl/^{203}Tl_{NIST\ 997}).$$

The terminology used for Tl isotope ratios is different from the standard δ-notation (variations in parts per 10,000 compared to 1,000, respectively). Generally, 2-sigma errors are reported for repeated dissolutions of standard reference materials.

The continental crust is the largest reservoir of Tl, as concentrations are high (Shaw, 1952; Nielsen et al., 2017). This is due to Tl being incompatible during mantle melting (Nielsen et al., 2014), as the depleted mantle has very low concentrations (Salters and Stracke, 2004). One direct measurement of a mantle xenolith has a value of −2.0 ± 0.8 (Nielsen et al., 2015), which is consistent with mid-ocean ridge basalt measurements of −2.0 ± 1.0 with very low concentrations (Nielsen et al., 2006a). Furthermore, samples of variable igneous origins, but importantly devoid of sedimentary inputs, have been analyzed and the combined data suggest the continental crust and bulk mantle Tl isotope values are indistinguishable, with $\varepsilon^{205}Tl$ values of −2.0 with a range of ± 1.0 (as reviewed in Nielsen et al., 2017). A recent study shows mineral separates from igneous and metamorphic rocks have a wide range of Tl isotope composition but many do cluster near −2.0, e.g., the whole rock value (Rader et al., 2018). Thus, fractionation from bulk Earth materials is unlikely because of near-quantitative transfer of Tl from the mantle to the crust.

Volcanic gases, riverine, hydrothermal fluid, and mineral aerosols (e.g., dust) are the dominant source fluxes into the ocean and are directly related to the mantle or weathering of the continental crust (Nielsen et al., 2005, 2006a, 2006b, 2007, 2016; Prytulak et al., 2013). The average Tl isotope compositions for these four sources are indistinguishable from those of the continental crust, as they all record $\varepsilon^{205}Tl$ values of −2.0, which represent ~83% of the total flux to the oceans (484.3 × 10^4 moles/yr; Rehkämper and Nielsen, 2004; Nielsen et al., 2005, 2006b, 2017; Baker et al., 2009). The last significant source flux is from reducing sedimentary pore fluids, which are currently estimated to be ~17% of the total input flux and have an isotopic value of 0 but generally display a larger isotopic range (Rehkämper and Nielsen, 2004; Nielsen et al., 2011b, 2017). The isotopic range is likely due to mixing, as Mn oxide dissolution might be captured in sulfidic pore fluids if these processes occur nearby, and/or due to nonquantitative Tl removal during pyrite formation (Owens et al., 2017). The isotopic sources of Tl isotopes are generally homogeneous. However, recent data document mineralogical variations (Rader et al., 2018), but this would require differential weathering and/or variable crustal mineralogy throughout Earth history to have a significant effect on the global marine Tl isotope composition.

The modern open-ocean seawater Tl isotope composition is homogeneous, with an ε^{205}Tl value of -6.0 ± 0.6 (Rehkämper et al., 2002; Nielsen et al., 2006b; Owens et al., 2017), with a relatively constant global concentration of 13.3 ppb (Flegal and Patterson, 1985; Rehkämper et al., 2002; Nielsen et al., 2006b; Owens et al., 2017). Thus, the modern marine isotopic composition is depleted (i.e., more negative) compared to that of the average continental crust and mantle. This is because the two dominant sinks of Tl, Mn oxide adsorption and incorporation during low-temperature alteration of oceanic crust (AOC; Nielsen et al., 2017), have large and opposing isotope compositions. The largest output flux is AOC, which accounts for ~65% of the Tl burial and has an average isotopic value of ~-7.5 (range between -6 and -12 ε units; Rehkämper and Nielsen, 2004; Nielsen et al., 2006b, 2017). Thus, the isotopic fractionation of AOC from modern seawater values is relatively small, with a difference of ~-1.5 ε units (e.g., the difference between AOC and seawater). However, there is a large isotopic range for AOC Tl isotopes related to the depth of the altered material, with greater depths exhibiting smaller fractionations (Nielsen et al., 2006b; Coggon et al., 2014). The mechanism controlling the incorporation of isotopically light Tl during alteration of oceanic crust remains ambiguous but could be kinetic in nature (Nielsen and Rehkämper, 2012b) or caused by microbially mediated sulfide precipitation (Coggon et al., 2014). The other significant Tl sink is associated with low-temperature authigenic Mn oxides (Rehkämper and Nielsen, 2004), specifically birnessite (Peacock and Moon, 2012). Manganese oxides have an average ε^{205}Tl composition of ~$+10$ with a range between $+6$ and $+12$, and Mn oxide preservation accounts for ~30% of the burial. High-temperature hydrothermal Mn oxide precipitates such as the mineral todorokite display no fractionation from seawater or hydrothermal fluids (Peacock and Moon, 2012; Nielsen et al., 2013). The average Mn oxide burial value of $+10$ provides an isotopic difference from seawater of ~$+16$. This suggests that small changes in the amount of Mn oxide burial can have a significant effect on the global seawater Tl isotope composition; thus, Tl isotopes can potentially track changes in the local or global burial flux of Mn oxides depending on the connectivity of a sedimentary basin with the open ocean.

There are two additional, albeit minor Tl sinks from the ocean and are estimated to represent 5% of the burial flux (Owens et al., 2017). They are associated with reducing bottom water conditions, specifically low oxygen (suboxic) and anoxic/euxinic settings. The fluxes and isotopic compositions for low oxygen environments are limited in the modern ocean. Limited data from reducing sediments suggest a range of values but an average near 0 (Nielsen et al., 2011b). This range of isotopic values in reducing sediments is likely due to Mn oxide reduction near sulfidic pore fluids; further research is needed, however, to constrain Tl behavior in these settings as a recent study suggests a potential to capture seawater Tl isotope values (Fan et al., in press). In addition, several estuaries or tidally influenced sediments have been analyzed for Tl concentrations (Böning et al., 2018 and references therein), but the global record remains unclear and lacks isotopic analysis. Euxinic sediments capture the oxic seawater value (Section 3) and the flux in these environments was estimated using Tl/S relationships in the Black Sea (Owens et al., 2017), but detailed research is needed to confirm the calculations of this flux.

There are no known enzymatic pathways that utilize and fractionate Tl. There are several known plant species, however, that accumulate thallium, likely because of the similar ionic radius of potassium and there is a fractionation within plants (Rader et al., 2019). With our current understanding, it is unlikely that biological cycling and perturbations to the marine biosphere have a significant effect on the oceanic Tl isotope mass balance. Furthermore, the modern vertical and horizontal ocean transects of Tl isotopes and concentrations in the South Atlantic document no observable fractionations, as the surface and deep waters record values that are analytically indistinguishable (Owens et al., 2017).

Therefore, temporal changes in the marine Tl isotope composition are most likely controlled by the ratio between the two marine output fluxes (Nielsen et al., 2009, 2011b; Owens et al., 2017), which are highly fractionated (Rehkämper et al., 2002; Rehkämper and Nielsen, 2004; Nielsen et al., 2006b, 2017). Importantly, AOC fluxes are likely globally controlled by ocean production rates driving oceanic crust cooling that change over long timescales – many millions of years. Therefore, short-term perturbations to the global marine Tl system are most likely driven by changes in the precipitation/dissolution, preservation, and burial flux of Mn oxides on the seafloor, a process that requires free oxygen. Thus, a decrease in the burial of the positively fractionated Mn oxides will shift the seawater composition toward more positive values.

3 Sedimentary Thallium Isotope Signature

Thallium exists in two valence states: Tl^+ and Tl^{3+}. The oxidized form is uncommon in natural environments because of the high redox potential, but it is likely that this form of Tl plays a significant role in the observed strong adsorption onto authigenic Mn oxides (Matthews and Riley, 1970; McGoldrick et al., 1979; Bidoglio et al., 1993; Hein et al., 2000; Peacock and Moon, 2012; Nielsen et al., 2013). The first studies to reconstruct ancient marine Tl isotope compositions analyzed ferro-manganese crusts (Rehkämper et al., 2004; Nielsen et al., 2009, 2011a); this assumes a constant and near modern isotopic fractionation in order to reconstruct seawater values (Rehkämper et al., 2002). There is currently no evidence to suggest this offset varies, systematically or otherwise, but the isotopic range for the Cenozoic era is between ~4 and 14 ε units. In addition, the utility of Fe–Mn crusts is limited because of the limited temporal preservation (maximum ages of ~70 million years) and age constraints (Nielsen et al., 2009). Thus, it was essential to develop an additional archive to increase the utility of Tl isotopes. However, a high-resolution ferro-manganese Tl isotope record has been generated for the Cenozoic era and provides evidence for large isotopic variations (Nielsen et al., 2009).

The first Tl isotope data analyzed from organic-rich sediments were generated from mechanically separated pyrite grains (Nielsen et al., 2011b), which provided the motivation to analyze Tl isotopes in environments that precipitated pyrite in the water column and thus had the potential to capture and record local and potentially global environmental conditions. A modern study of Tl isotopes of seawater and sediment core tops from the two largest modern euxinic basins, Cariaco Basin and the Black Sea, was conducted as a first assessment to reconstruct seawater values (Owens et al., 2017). The overlying oxic seawater values in the Cariaco Basin values are −5. 6 ± 0.7, which are within analytical uncertainty of open ocean values, while the Black Sea values are −2.2 ± 0.3 (one sample), which are similar to input fluxes (Owens et al., 2017). Core top samples from the Cariaco Basin documented an average $\varepsilon^{205}Tl_{leach}$ −5.4 ± 0.6 for five samples (excluding one sample), while the Black Sea documented average values of −2.3 ± 0.6 for the leached fraction of five samples. This initial research documented that the leached fraction (see Section 5 for more details) of the sedimentary core tops from both basins captures, within analytical uncertainties, the overlying oxic seawater value from each basin. The Tl isotope difference for each locality is likely due to basin restriction. The Cariaco Basin is considered a more open-ocean archive and the Black Sea is severely restricted, and when coupled to the relatively short residence time of Tl allows for isotopic variation between these two basins. In addition, recent work highlights the utilization of anoxic but noneuxinic sediments to capture seawater values (Fan et al., in press) from the Santa Barbara Basin, where sulfide is limited to the pore fluids. Limited data from reducing sediments overlain by an oxic water-column do no capture seawater values (Nielsen et al., 2011b). The difference in these depositional environments suggests that the distance between Mn oxide dissolution and pyrite formation is important and needs to be investigated in greater detail (see Section 6 for more discussion).

Moreover, Tl concentrations below the chemocline (euxinic) decrease rapidly with depth (Owens et al., 2017). This drawdown of Tl is not likely due to direct Tl sulfide precipitation, as the solubility in aqueous solution is high (Nielsen et al., 2011b). More likely, it reflects Tl partitioning into Fe sulfides that precipitate in the water column, as Tl is known to have a strong affinity for such minerals (McGoldrick et al., 1979; Nielsen et al., 2011b, 2014). The near-quantitative removal of Tl from euxinic seawater (Owens et al., 2017) likely accounts for the indistinguishable Tl isotope compositions observed for sediments deposited under euxinic conditions and the overlying oxic waters. Thus, it is important to constrain local redox conditions using other proxies that are not susceptible to global redox perturbations, such as Fe speciation, Mn concentrations, or other evidence for local oxide-bearing minerals (Owens et al., 2016; Ostrander et al., 2017; Them et al., 2018). Most significantly, this work documents the ability of sediments deposited under euxinic and, potentially ferruginous, conditions to capture the overlying oxic seawater values. When possible, sediments from multiple ocean basins should be analyzed to confidently capture global seawater perturbations (Ostrander et al., 2017; Them et al., 2018). This is not always possible, however, owing to biases in the preservation of strata in the geological record; thus, invariant stratigraphic trends should be interpreted with caution, as they could be due to basin restriction (i.e., modern Black Sea) or coupled with additional proxies.

4 Brief Analytical Guide

An operationally defined chemical separation (leaching method) has been developed using modern euxinic sediments to extract the oxic seawater value. This was required to separate the authigenic Tl (likely complexed with pyrite) from the lithogenic (likely bound to silicates) fraction, as the bulk values did not capture oxic seawater value but rather a mixture of the two isotopic pools (Owens et al., 2017). This is classically observed in the Cariaco Basin, as the bulk sediment values are -2.5 ε units, while the authigenic values are -5.5 ε units and ~63% of the total Tl concentration and the lithogenic values are -2.0 ε units and 37% (Owens et al., 2017). This relationship is not observed in the Black Sea, however, as the lithogenic and oxic seawater are nearly the same (i.e., within analytical error; Owens et al., 2017) but the percentage of authigenic to total Tl is similar to that of the Cariaco Basin (Owens et al., 2017). Authigenic concentrations of ancient marine sediments have been greater than the modern values, and therefore the bulk values would be less affected by lithogenic contamination, but the Tl leach method is still required to compare similar Tl isotope fractions. For detailed information on the method, refer to Owens et al. (2017). The chemically separated solutions are then purified using an established anion-exchange protocol (Rehkämper and Halliday, 1999; Rehkämper et al., 2002; Nielsen et al., 2005, 2009; Baker et al., 2009; Nielsen and Rehkämper, 2012a). Ostrander et al. (2017) documented that only a single microcolumn is required for samples with high Tl to Pb ratios. The purified solution is then analyzed using a multicollector–inductively coupled plasma–mass spectrometer (MC-ICP-MS). The leaching method on the USGS SCo-1 standard has an average ε^{205}Tl of −2.99 with a long-term reproducibility better than 0.3 ε units (Ostrander et al., 2017, 2019; Owens et al., 2017; Them et al., 2018), which is within or slightly better than bulk ε^{205}Tl values (Nielsen et al., 2014).

5 Case Studies

While the proxy is in its infancy and there are many new avenues to explore, there have been several applications of Tl isotopes in ancient marine shales to reconstruct the global burial of Mn oxides to track the initial marine oxygen perturbation. Nielsen et al. (2009) reconstructed the Tl isotope record for the Cenozoic era by analyzing Mn oxide crusts, which documents a secular trend to heavier values over the past ~60 Myr. The nearly 10-ε unit shift between 55 and 45 million years has been interpreted as an increased Mn oxide burial, as there is no evidence for changes in AOC because spreading rates did not change dramatically (Nielsen et al., 2009). Ideally, this will be confirmed using additional archives to constrain the global seawater record and constrain the fractionations associated with Mn oxide burial. In addition, the combination of Tl isotopes and euxinic proxies (i.e., molybdenum isotopes and concentrations, carbonate-associated sulfur isotopes, etc.) have the potential to delineate post-oxic (non-sulfidic anoxia) and euxinic conditions. Thus, a more holistic reconstruction of the global redox landscape is possible using a combination of redox proxies to illuminate the timing and magnitude of redox perturbations. It is important to constrain the relationship between ancient redox conditions and volcanism, carbon isotope excursions, and extinction events in the geological record (Fig. 3).

The geologically youngest application of organic-rich shales that capture seawater Tl isotopes to date has been across Cretaceous Oceanic Anoxic Event 2 at the Cenomanian/Turonian Boundary Event (~94 Ma). This event has been well studied with organic-rich mudstone deposition recorded from multiple ocean basins and a documented redox perturbation (Owens et al., 2018

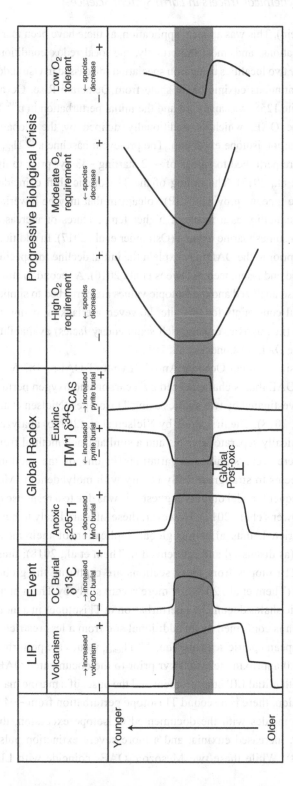

Figure 3 Idealized relative stratigraphic perturbations of volcanism from large igneous provinces, carbon isotopes ($\delta^{13}C$) documenting organic carbon (OC) burial, global redox conditions using anoxic (Tl isotopes) and euxinic proxies, and the known stepwise extinction for various events. Note global post-oxic conditions can be inferred using the stratigraphic offset and the magnitude of seafloor anoxia ($\epsilon^{205}Tl$) and euxinia.

and references therein). This was an ideal application, as there have been many geochemical applications, and, most importantly, the local redox conditions were constrained for two localities using iron speciation (discussed in Ostrander et al., 2017). A continuous euxinic black shale from Demerara Rise, Ocean Drilling Program Site 1258, was analyzed and the initial perturbation in $\varepsilon^{205}Tl$ values precedes the OAE, which is traditionally defined by the globally observed positive carbon isotope excursion. The pre-event baseline $\varepsilon^{205}Tl_{leach}$ values of ~−4.5 were perturbed to values of ~−2 starting ~43 kyr prior to the event (Ostrander et al., 2017). The timing of the Tl isotopic shift coincides with increased large igneous province (LIP) volcanism that may be lowering gas (i.e., oxygen) solubility as a result of higher temperatures or increased productivity that consumes marine oxygen (Ostrander et al., 2017). In addition, this deoxygenation prior to the OAE may explain the initial decline for specific species (as discussed and referenced in Owens et al., 2016). A second locality, Furlo (Italy), was also analyzed and the isotopic values are perturbed to similar values; the temporal constraints for this site, however, are lacking owing to limited black shale (i.e., anoxic, organic-rich sedimentary facies) availability prior to and after the OAE (Ostrander et al., 2017).

The Early Jurassic Toarcian Oceanic Anoxic Event (T-OAE; ~183 Ma), another Mesozoic OAE that is characterized by a carbon and oxygen perturbation, has been investigated by two studies using Tl isotopes (Nielsen et al., 2011b; Them et al., 2018). The first study by Nielsen et al. (2011b) analyzed Tl isotopes of physically separated pyrites and a similar leach method from two localities. There were significant variations in one of the sections analyzed, which seems to stratigraphically covary with molybdenum (Mo) isotopes, and subsequent model outputs suggest this was due to a response to Mn oxide burial (Owens et al., 2017). However, these sites are likely to have been severely restricted, thus affecting global application of their metal isotope variations (as discussed and referenced in Them et al., 2018); thus, it is unlikely the Tl isotopes from these sections are capturing the global seawater signature (Them et al., 2018). A more recent study by Them et al. (2018) using euxinic shales document a perturbation in Tl isotopes in a more open locality, which is confirmed by an additional site from a less restricted portion of the European epeiric seaway. The $\varepsilon^{205}Tl_{leach}$ values are perturbed from ~−6.0 to ~−4.0 approximately 600 kyr prior to the documented OAE, which coincides with initial LIP emplacement and the onset of a marine mass extinction. In addition, there is a second Tl isotope perturbation from ~−4.0 to ~−2.0, which coincides with the documented C isotope excursion, the main phase of LIP, increased euxinia, and a more severe extinction pulse (Them et al., 2018). While these two Mesozoic OAEs coincide with LIP emplacement it is unlikely that all Tl isotope perturbations are driven by volcanism.

The combination of Tl isotopes and carbonate-associated sulfate (CAS) sulfur isotopes can help to delineate global post-oxic and euxinic conditions as CAS S-isotopes respond to global pyrite burial, which is most efficient under euxinic conditions. Thus, intervals that display Tl isotope perturbations but lack concurrent CAS perturbations are likely indicative of a global increase in post-oxic conditions, which has been observed prior to the carbon isotope perturbations for OAE 2 and T-OAE. The return to pre-event Tl isotopes is not observed for either Mesozoic OAE, which has been suggested to be related to continued low oxygen bottom waters due to previously buried organic carbon that prevented Mn oxide burial (Them et al., 2018). Authigenic Tl concentrations have not shown any systematic variations across these two OAEs (Ostrander et al., 2017; Them et al., 2018), which suggests the marine inventory is not depleted due to increased anoxic conditions, unlike other redox-sensitive trace metals during these periods (Owens et al., 2016). Current research thus suggests the residence time of Tl is not dramatically perturbed during these events.

The third published shale Tl isotope study to date added Mo isotopes to better constrain the evidence for marine water-column oxygenation which required free oxygen to penetrate the sediments prior to the Great Oxidation Event, ~2.5 Ga (Ostrander et al., 2019). This evidence suggests that oxygenated continental shelves predated the local evidence for oxidative continental weathering in the same samples (Ostrander et al., 2019). This study documents a small but resolvable perturbation in both isotopic systems that generally covaries with a positive ~2-ε unit shift in $\varepsilon^{205}Tl_{leach}$ (−1.6 to −3.6) and a negative ~1 ‰ shift in Mo isotopes (1.5 to 0.5), which suggests each system is responding to a similar perturbation. The largest isotopic leverage that exists to drive both systems is manganese oxides, as they require free oxygen to persist through the water column and sediments for Mn burial. For this particular study, combining Mo and Tl isotopes helped to constrain the mechanism driving these perturbations, as there are fractionations associated with both systems.

6 Future Directions

The current research and application of Tl isotopes throughout the geological record have provided new insights into the earliest changes of ocean oxygenation. While the proxy is still in the developmental stage, there seems to be great progress and promise for the application in a wide array of Earth System–related topics. The low-temperature surficial application of Tl is a new avenue of exploration. There is a need for more detailed studies investigating various depositional environments for the fluxes and their respective isotopic signatures, as well as potential new sources. Understanding how individual mineral variations (Rader et al., 2018) could change throughout Earth's history and/or weathering regimes will need to be further investigated, but they are unlikely to have a significant effect on short-term Tl isotopic perturbations.

Given the wide isotopic range for the two largest burial fluxes, AOC and Mn oxides, there is a need to understand the mechanisms related to the rates and associated fractionations. Thus, investigations that use laboratory and natural systems are needed to better constrain the fluxes and isotopic fractionations. This is required in order to increase the confidence of quantitative estimates of manganese oxide deposition to better constrain variations in marine oxic conditions during ancient climate events. In addition, Tl concentrations and isotope signatures need to be further investigated in areas of low-oxygen bottom waters and sediments, which includes regions of the ocean that locally bury Mn oxides initially but are then remineralized during early diagenesis in environments that contain pore water sulfide. These environments, however, are likely important for Mo mass balance relative to Mn oxide cycling (Reinhard et al., 2013). Thus, it is necessary to better constrain Mn cycling and any related Tl isotope variations to further refine and develop the proxy.

Using a multiproxy approach has the potential to better constrain the input and outputs of the geochemical system, especially with those proxies that have similar and overlapping but unconstrained parameters. For example, combining a euxinic proxy with Tl isotopes has the potential to delineate the spatiotemporal record for post-oxic anoxia and euxinic anoxia (Them et al., 2018), or combining Mo and Tl isotopes to better constrain the burial fluxes of manganese oxides (Ostrander et al., 2019). In addition, Tl isotopes may have the potential to track basin restriction using multiple localities and stratigraphic constraints to compare isotopic signatures. It is likely there are numerous unexplored applications related to sedimentary Tl isotopes.

7 Key Papers

Isotopic Measurement

Initial papers that document the thallium purification using ion-exchange chromatography and analytical analysis.

Rehkämper, M., Halliday, A.N., 1999. The precise measurement of Tl isotopic compositions by MC-ICPMS: Application to the analysis of geological materials and meteorites. *Geochimica et Cosmochimica Acta* 63, 935–944.

Rehkämper, M., Frank, M., Hein, J.R., Porcelli, D., Halliday, A., Ingri, J., Liebetrau, V., 2002. Thallium isotope variations in seawater and hydrogenetic, diagenetic, and hydrothermal ferromanganese deposits. *Earth and Planetary Science Letters* 197, 65–81.

Reviews and marine mass balance

Two papers that provide the development and understanding of the thallium isotope marine mass balance.

Rehkämper, M., Nielsen, S.G., 2004. The mass balance of dissolved thallium in the oceans. Marine Chemistry 85, 125–139.

Nielsen, S.G., Rehkämper, M., Prytulak, J., 2017. Investigation and Application of Thallium Isotope Fractionation. Reviews in Mineralogy & Geochemistry 82, 759–798.

Isotope fractionations of the major marine sinks

Compares the isotopic range of hydrogenatic ferromanganese crusts to other sedimentary archives to interpret the marine record of Tl isotopic variability.

Rehkämper, M., Frank, M., Hein, J.R., Halliday, A., 2004. Cenozoic marine geochemistry of thallium deduced from isotopic studies of ferromanganese crusts and pelagic sediments. Earth and Planetary Science Letters 219, 77–91.

Provides evidence that birnessite, an Mn oxide, can oxidatively scavenge thallium and imparts a positive thallium isotope fractionation, unlike other Mn oxides.

Peacock, C.L., Moon, E.M., 2012. Oxidative scavenging of thallium by birnessite: Explanation for thallium enrichment and stable isotope fractionation in marine ferromanganese precipitates. Geochimica et Cosmochimica Acta 84, 297–313.

This manuscript provides further evidence that birnessite imparts a large positive thallium isotope during various sorption experiments.

Nielsen, S.G., Wasylenki, L.E., Rehkämper, M., Peacock, C.L., Xue, Z., Moon, E.M., 2013. Towards an understanding of thallium isotope fractionation during adsorption to manganese oxides. Geochimica et Cosmochimica Acta 117, 252–265.

Provides details on the limited isotopic fractionation in altered oceanic crust but an observed variation with depth.

Coggon, R.M., Rehkämper, M., Atteck, C., Teagle, D.A.H., Alt, J.C., Cooper, M.J., 2014. Controls on thallium uptake during hydrothermal alteration of the upper ocean crust. Geochimica et Cosmochimica Acta 144, 25–42.

Organic-rich sediments tracking modern seawater

This manuscript provides the framework for analyzing organic-rich sediments to capture oxic seawater values.

Owens, J.D., Nielsen, S.G., Horner, T.J., Ostrander, C.M., Peterson, L.C., 2017. Thallium-isotopic compositions of euxinic sediments as a proxy for global manganese-oxide burial. Geochimica et Cosmochimica Acta 213, 291–307.

Ancient applications

Uses ferro-manganese crusts through the Cenozoic to infer global variations in the marine Tl isotope record.

Nielsen, S.G., Mar-Gerrison, S., Gannoun, A., LaRowe, D., Klemm, V., Halliday, A.N., Burton, K.W., Hein, J.R., 2009. Thallium isotope evidence for a permanent increase in marine organic carbon export in the early Eocene. Earth and Planetary Science Letters 278, 297–307.

This paper utilizes Tl isotopes leached from organic-rich sediments to capture the global marine perturbation to estimate the extent of anoxia.

Ostrander, C.M., Owens, J.D., Nielsen, S.G., 2017. Constraining the rate of oceanic deoxygenation leading up to a Cretaceous Oceanic Anoxic Event (OAE-2: ~94 Ma). Science Advances 3.

Analysis of Tl isotopes leached from organic-rich open ocean marine sediments to record the global record of anoxia during the Toarcian Oceanic Anoxic Event.

Them, T.R., Gill, B.C., Caruthers, A.H., Gerhardt, A.M., Gröcke, D.R., Lyons, T.W., Marroquín, S.M., Nielsen, S.G., Trabucho Alexandre, J.P., Owens, J.D., 2018. Thallium isotopes reveal protracted anoxia during the Toarcian (Early Jurassic) associated with volcanism, carbon burial, and mass extinction. Proceedings of the National Academy of Sciences 115, 6596.

This paper utilizes Tl isotopes coupled with molybdenum isotopes from organic-rich black shales to interpret an increase in burial of Mn oxides thus an increase in oxygen.

Ostrander, C.M., Nielsen, S.G., Owens, J.D., Kendall, B., Gordon, G.W., Romaniello, S.J., Anbar, A.D., 2019. Fully oxygenated water columns over continental shelves before the Great Oxidation Event. Nature Geoscience 12, 186–191.

References

Anbar, A. D., Knoll, A. H. 2002. Proterozoic ocean chemistry and evolution: A bioinorganic bridge? *Nature* 297, 1137–1142.

Anbar, A. D., Rouxel, O. 2007. Metal stable isotopes in paleoceanography. *Annual Review of Earth and Planetary Sciences* 35, 717–746.

Baker, R. G. A., Rehkämper, M., Hinkley, T. K., Nielsen, S. G., Toutain, J. P. 2009. Investigation of thallium fluxes from subaerial volcanism: Implications for the present and past mass balance of thallium in the oceans. *Geochimica et Cosmochimica Acta* 73, 6340–6359.

Berner, R. A. 1981. A new geochemical classification of sedimentary environments. *Journal of Sedimentary Research* 51, 359–365.

Berner, R. A. 2006. GEOCARBSULF: A combined model for Phanerozoic atmospheric O_2 and CO_2. *Geochimica et Cosmochimica Acta: A Special Issue Dedicated to Robert A. Berner* 70, 5653–5664.

Bidoglio, G., Gibson, P. N., O'Gorman, M., Roberts, K. J. 1993. X-ray absorption spectroscopy investigation of surface redox transformations of thallium and chromium on colloidal mineral oxides. *Geochimica et Cosmochimica Acta* 57, 2389–2394.

Böning, P., Schnetger, B., Beck, M., Brumsack, H.-J. 2018. Thallium dynamics in the southern North Sea. *Geochimica et Cosmochimica Acta* 227, 143–155.

Canfield, D. E., Thamdrup, B. 2009. Towards a consistent classification scheme for geochemical environments, or, why we wish the term 'suboxic' would go away. *Geobiology* 7, 385–392.

Coggon, R. M., Rehkämper, M., Atteck, C., Teagle, D. A. H., Alt, J. C., Cooper, M. J. 2014. Controls on thallium uptake during hydrothermal alteration of the upper ocean crust. *Geochimica et Cosmochimica Acta* 144, 25–42.

Fan, H., Nielsen, S. G., Owens, J. D., Auro, M., Shu, Y., Hardisty, D. S., Bowman, C., Young, S. A., Wen, H. In press. Constraining oceanic oxygenation during the Shuram excursion in South China using thallium isotopes. *Geobiology*.

Flegal, A. R., Patterson, C. C. 1985. Thallium concentrations in seawater. *Marine Chemistry* 15, 327–331.

Froelich, P. N., Klinkhammer, G. P., Bender, M. L., et al. 1979. Early oxidation of organic matter in pelagic sediments of the eastern equatorial Atlantic: Suboxic diagenesis. *Geochimica et Cosmochimica Acta* 43, 1075–1090.

Glass, J.B., Wolfe-Simon, F., Anbar, A.D. 2009. Coevolution of metal availability and nitrogen assimilation in cyanobacteria and algae. *Geobiology* 7, 100–123.

Hannisdal, B., Peters, S. E. 2011. Phanerozoic Earth System evolution and marine biodiversity. *Science* 334, 1121–1124.

Hardisty, D. S., Lyons, T. W., Riedinger, N., et al. 2018. An evaluation of sedimentary molybdenum and iron as proxies for pore fluid paleoredox conditions. *American Journal of Science* 318, 527–556.

Hein, J. R., Koschinsky, A., Bau, M., Manheim, F. T., Kang, J.-K., Roberts, L. 2000. Cobalt-rich ferromanganese crusts in the Pacific. *Handbook of Marine Mineral Deposits* 18, 239–273.

Lyons, T. W., Reinhard, C. T., Planavsky, N. J. 2014. The rise of oxygen in Earth's early ocean and atmosphere. *Nature* 506, 307–315.

Matthews, A. D., Riley, J. P. 1970. The occurrence of thallium in sea water and marine sediments. *Chemical Geology* 6, 149–152.

McGoldrick, P. J., Keays, R. R., Scott, B. B. 1979. Thallium: A sensitive indicator of rock/seawater interaction and of sulfur saturation of silicate melts. *Geochimica et Cosmochimica Acta* 43, 1303–1311.

Nielsen, S., Rehkämper, M. 2012a. Thallium isotopes and their application to problems in Earth and environmental science. In M. Baskaran (ed.), *Handbook of Environmental Isotope Geochemistry*. Berlin and Heidelberg: Springer, pp. 247–269.

Nielsen, S. G., Gannoun, A., Marnham, C., Burton, K. W., Halliday, A. N., Hein, J. R. 2011a. New age for ferromanganese crust 109D-C and implications for isotopic records of lead, neodymium, hafnium, and thallium in the Pliocene Indian Ocean. *Paleoceanography* 26, PA2213.

Nielsen, S. G., Goff, M., Hesselbo, S. P., Jenkyns, H. C., LaRowe, D. E., Lee, C.-T. A. 2011b. Thallium isotopes in early diagenetic pyrite: A paleoredox proxy? *Geochimica et Cosmochimica Acta* 75, 6690–6704.

Nielsen, S. G., Klein, F., Kading, T., Blusztajn, J., Wickham, K. 2015. Thallium as a tracer of fluid–rock interaction in the shallow Mariana forearc. *Earth and Planetary Science Letters* 430, 416–426.

Nielsen, S. G., Mar-Gerrison, S., Gannoun, A., et al. 2009. Thallium isotope evidence for a permanent increase in marine organic carbon export in the early Eocene. *Earth and Planetary Science Letters* 278, 297–307.

Nielsen, S. G., Rehkämper, M. 2012b. Thallium isotopes and their application to problems in earth and environmental science. In M. Baskaran (ed.), *Handbook of Environmental Isotope Geochemistry*. Berlin and Heidelberg: Springer, pp. 247–269.

Nielsen, S. G., Rehkämper, M., Baker, J., Halliday, A. N. 2004. The precise and accurate determination of thallium isotope compositions and concentrations for water samples by MC-ICPMS. *Chemical Geology* 204, 109–124.

Nielsen, S. G., Rehkämper, M., Brandon, A. D., Norman, M. D., Turner, S., O'Reilly, S. Y. 2007. Thallium isotopes in Iceland and Azores lavas: Implications for the role of altered crust and mantle geochemistry. *Earth and Planetary Science Letters* 264, 332–345.

Nielsen, S. G., Rehkamper, M., Norman, M. D., Halliday, A. N., Harrison, D. 2006a. Thallium isotopic evidence for ferromanganese sediments in the mantle source of Hawaiian basalts. *Nature* 439, 314–317.

Nielsen, S. G., Rehkämper, M., Porcelli, D., et al. 2005. Thallium isotope composition of the upper continental crust and rivers: An investigation of the continental sources of dissolved marine thallium. *Geochimica et Cosmochimica Acta* 69, 2007–2019.

Nielsen, S. G., Rehkämper, M., Prytulak, J. 2017. Investigation and application of thallium isotope fractionation. *Reviews in Mineralogy & Geochemistry* 82, 759–798.

Nielsen, S. G., Rehkämper, M., Teagle, D. A. H., Butterfield, D. A., Alt, J. C., Halliday, A. N. 2006b. Hydrothermal fluid fluxes calculated from the isotopic mass balance of thallium in the ocean crust. *Earth and Planetary Science Letters* 251, 120–133.

Nielsen, S. G., Shimizu, N., Lee, C.-T. A., Behn, M. D. 2014. Chalcophile behavior of thallium during MORB melting and implications for the sulfur content of the mantle. *Geochemistry, Geophysics, Geosystems* 15, 4905–4919.

Nielsen, S. G., Wasylenki, L. E., Rehkämper, M., Peacock, C. L., Xue, Z., Moon, E. M. 2013. Towards an understanding of thallium isotope fractionation during adsorption to manganese oxides. *Geochimica et Cosmochimica Acta* 117, 252–265.

Nielsen, S. G., Yogodzinski, G., Prytulak, J., et al. 2016. Tracking along-arc sediment inputs to the Aleutian arc using thallium isotopes. *Geochimica et Cosmochimica Acta* 181, 217–237.

Ostrander, C. M., Nielsen, S. G., Owens, J. D., et al. 2019. Fully oxygenated water columns over continental shelves before the Great Oxidation Event. *Nature Geoscience* 12, 186–191.

Ostrander, C. M., Owens, J. D. and Nielsen, S. G., 2017. Constraining the rate of oceanic deoxygenation leading up to a Cretaceous Oceanic Anoxic Event (OAE-2: ~94 Ma). *Science advances*, 3(8), p.e1701020.

Owens, J. D., Lyons, T. W., Lowery, C. M. 2018. Quantifying the missing sink for global organic carbon burial during a Cretaceous oceanic anoxic event. *Earth and Planetary Science Letters* 499, 83–94.

Owens, J. D., Nielsen, S. G., Horner, T. J., Ostrander, C. M., Peterson, L. C. 2017. Thallium-isotopic compositions of euxinic sediments as a proxy for global manganese-oxide burial. *Geochimica et Cosmochimica Acta* 213, 291–307.

Owens, J. D., Reinhard, C. T., Rohrssen, M., Love, G. D., Lyons, T. W. 2016. Empirical links between trace metal cycling and marine microbial ecology during a large perturbation to Earth's carbon cycle. *Earth and Planetary Science Letters* 449, 407–417.

Peacock, C. L., Moon, E. M. 2012. Oxidative scavenging of thallium by birnessite: Explanation for thallium enrichment and stable isotope fractionation in marine ferromanganese precipitates. *Geochimica et Cosmochimica Acta* 84, 297–313.

Prytulak, J., Nielsen, S., Plank, T., Barker, M., Elliott, T. 2013. Assessing the utility of thallium and thallium isotopes for tracing subduction zone inputs to the Mariana arc. *Chemical Geology* 345, 139–149.

Rader, S. T., Maier, R. M., Barton, M. D. and Mazdab, F. K., 2019. Uptake and Fractionation of Thallium by Brassica juncea in a Geogenic Thallium-Amended Substrate. *Environmental science & technology*, 53(5), pp. 2441–2449.

Rader, S. T., Mazdab, F. K., Barton, M. D. 2018. Mineralogical thallium geochemistry and isotope variations from igneous, metamorphic, and metasomatic systems. *Geochimica et Cosmochimica Acta* 243, 42–65.

Raiswell, R., Hardisty, D. S., Lyons, T. W., et al. 2018. The iron paleoredox proxies: A guide to the pitfalls, problems and proper practice. *American Journal of Science* 318, 491–526.

Rehkämper, M., Frank, M., Hein, J. R., Halliday, A. 2004. Cenozoic marine geochemistry of thallium deduced from isotopic studies of ferromanganese crusts and pelagic sediments. *Earth and Planetary Science Letters* 219, 77–91.

Rehkämper, M., Frank, M., Hein, J. R., et al. 2002. Thallium isotope variations in seawater and hydrogenetic, diagenetic, and hydrothermal ferromanganese deposits. *Earth and Planetary Science Letters* 197, 65–81.

Rehkämper, M., Halliday, A. N. 1999. The precise measurement of Tl isotopic compositions by MC-ICPMS: Application to the analysis of geological materials and meteorites. *Geochimica et Cosmochimica Acta* 63, 935–944.

Rehkämper, M., Nielsen, S. G. 2004. The mass balance of dissolved thallium in the oceans. *Marine Chemistry* 85, 125–139.

Reinhard, C.T., Planavsky, N.J., Robbins, L.J., et al. 2013. Proterozoic ocean redox and biogeochemical stasis. *Proceedings of the National Academy of Sciences of the USA* 110, 5357–5362.

Rue, E. L., Smith, G. J., Cutter, G. A., Bruland, K. W. 1997. The response of trace element redox couples to suboxic conditions in the water column. *Deep Sea Research Part I: Oceanographic Research Papers* 44, 113–134.

Saito, M. A., Goepfert, T. J., Ritt, J. T. 2008. Some thoughts on the concept of colimitation: Three definitions and the importance of bioavailability. *Limnology and Oceanography* 53, 276–290.

Salters, V. J. M., and Stracke, A. (2004), Composition of the depleted mantle, *Geochem. Geophys. Geosyst.*, 5, Q05B07, doi: 10.1029/2003GC000597.

Saltzman, M. R. and Thomas, E., Carbon isotope stratigraphy, in The Geologic Time Scale, vol. 1, Elsevier, 2012, pp. 207–232.

Scott, C., Lyons, T. W. 2012. Contrasting molybdenum cycling and isotopic properties in euxinic versus non-euxinic sediments and sedimentary rocks: Refining the paleoproxies. *Chemical Geology* 324–325, 19–27.

Scott, C., Lyons, T. W., Bekker, A., et al. 2008. Tracing the stepwise oxygenation of the Proterozoic ocean. *Nature* 452, 456–459.

Shaw, D. M. 1952. The geochemistry of thallium. *Geochimica et Cosmochimica Acta* 2, 118–154.

Them, T. R., Gill, B. C., Caruthers, A. H., et al. 2018. Thallium isotopes reveal protracted anoxia during the Toarcian (Early Jurassic) associated with volcanism, carbon burial, and mass extinction. *Proceedings of the National Academy of Sciences of the USA* 115, 6596.

Tribovillard, N., Algeo, T. J., Lyons, T., Riboulleau, A. 2006. Trace metals as paleoredox and paleoproductivity proxies: An update. *Chemical Geology* 232, 12–32.

Xiong, Y. 2007. Hydrothermal thallium mineralization up to 300 C: A thermodynamic approach. *Ore Geology Reviews* 32, 291–313.

Acknowledgments

I thank T. R. Them for discussions and edits on an early draft of this chapter. The laboratory group at Florida State University and Woods Hole Oceanographic Institution and the many collaborators who have been working on thallium research. Two anonymous reviewers have focused this review and Tim Lyons has provided editorial guidance. This work has been funded by the National Science Foundation (OCE-1624895 and DMR-1157490), and NASA Exobiology (NNX16AJ60G and 80NSSC18K1532).

Elements in Geochemical Tracers in Earth System Science

Timothy Lyons

University of California, Riverside

Timothy Lyons is Distinguished Professor of Biogeochemistry in the Department of Earth Sciences at the University of California, Riverside. He is an expert in the use of geochemical tracers for applications in astrobiology, geobiology, and Earth history. Professor Lyons leads the Alternative Earths team of the NASA Astrobiology Institute and the Alternative Earths Astrobiology Center at UC Riverside.

Alexandra Turchyn

University of Cambridge

Alexandra Turchyn is a university reader in biogeochemistry in the Department of Earth Sciences at the University of Cambridge. Her primary research interests are in isotope geochemistry and the application of geochemical systems to interrogate modern and past environments.

Chris Reinhard

Georgia Institute of Technology

Chris Reinhard is an assistant professor in the Department of Earth and Atmospheric Sciences at the Georgia Institute of Technology. His research focuses on biogeochemistry and paleoclimatology, and he is an Institutional PI on the Alternative Earths team of the NASA Astrobiology Institute.

About the series

This innovative series provides authoritative, concise overviews of the many novel isotope and elemental systems that can be used as 'proxies' or 'geochemical tracers' to reconstruct past environments over thousands to millions to billions of years – from the evolving chemistry of the atmosphere and oceans to their cause-and-effect relationships with life.

Covering a wide variety of geochemical tracers, the series reviews each method in terms of the geochemical underpinnings, the promises and pitfalls, and the 'state-of-the-art' and future prospects, providing a dynamic reference resource for graduate students and researchers in geochemistry, astrobiology, paleontology, paleoceanography, and paleoclimatology.

The short, timely, broadly accessible papers provide much-needed primers for a wide audience – highlighting the cutting-edge of both new and established proxies as applied to diverse questions about Earth system evolution over wide-ranging time scales.

Cambridge Elements ≡

Elements in Geochemical Tracers in Earth System Science

Printed in the United States
By Bookmasters